# THE SOLAR SYSTEM

**MAURA GOUCK**

THE CHILD'S WORLD

**DESIGN**
Michael George

**PHOTO CREDITS**
NASA
Hale Observatories
Lick Observatory
National Optical Astronomy Observatories
Palomar Observatory

Text Copyright © 1993 by The Child's World, Inc.
All rights reserved. No part of this book may be
reproduced or utilized in any form or by any means
without written permission from the publisher.
Printed in the United States of America.

Library of Congress Cataloging-in-Publication Data
available upon request.
ISBN 1-56766-061-4

Distributed to schools and libraries in the United States by
ENCYCLOPAEDIA BRITANNICA EDUCATIONAL CORP.
310 South Michigan Avenue
Chicago, Illinois 60604

On a clear, moonless night the sky seems to be filled with stars. Watch carefully, and you'll see that these twinkling stars are different colors—yellow, red, bluish white. You'll probably spot the *Big Dipper*, *Orion*, or another group of stars. If you are lucky, you might even see a meteor streaking across the sky!

As daylight arrives, the night stars fade from view. At sunrise there is only one star that you can still see. It is the star nearest to our planet. This star is the *Sun*. The Sun is very important to us because it provides light, heat, and energy for our planet.

---

⇐

*The Sun and other stars of the Milky Way galaxy*

To us, the Sun seems to be about the same size as the Moon, but it is really much larger. In fact, the Sun is about four hundred times wider than the Moon. The Sun is also much larger than our own planet Earth. It would take over one million Earths to fill the Sun!

You can't stare at the Sun the way you stare at other stars. The Sun is so bright it hurts your eyes. Scientists look at the Sun through special instruments. They see a lot happening on this star! To us the Sun looks like a peaceful, yellow disk floating slowly across the sky. Actually, though, it is a ball of bubbling, churning gases. Powerful storms swirl across its surface, and eruptions of hot gases shoot thousands of miles into space.

---

⇒
*The Sun and planet Earth*

The Sun is the center of our *solar system*. The solar system consists of the Sun and the planets, asteroids, and comets that circle it. There are nine planets in our solar system. Each one follows a different path, or *orbit*, around the Sun.

The planet closest to the Sun is *Mercury*. It is a small planet, less than half the size of Earth. Mercury has a hard, rocky surface covered by many *craters*. Craters are circular holes formed when rocks, called *asteroids*, crashed into the planet. The temperature on Mercury ranges from very hot to icy cold. Humans could not survive on this planet because it has no water and very little air.

The hottest planet is *Venus*, the second planet from the Sun. The temperature on Venus is

---

⇐
*Mercury's horizon and its cratered surface*

hotter than a pizza oven! Venus is about the same size as Earth. Sometimes these two are called sister planets, but they really have very little in common. Venus is covered by thick, yellow clouds—but they're not rain clouds. They're made of poisonous acid! Venus and Earth even spin in opposite directions. If you lived on Venus, you would watch the Sun rise in the west and set in the east!

We live on *Earth*, the third planet from the Sun. Earth is about 93 million miles from the Sun. It would take more than fifteen years to fly from Earth to the Sun in a jet plane!

Photographs taken from space show us that Earth is a beautiful blue ball surrounded by a layer of air—our *atmosphere*. The blue, of course, is

---

⇒
*A computer-enhanced photo shows hills and valleys on the surface of Venus*

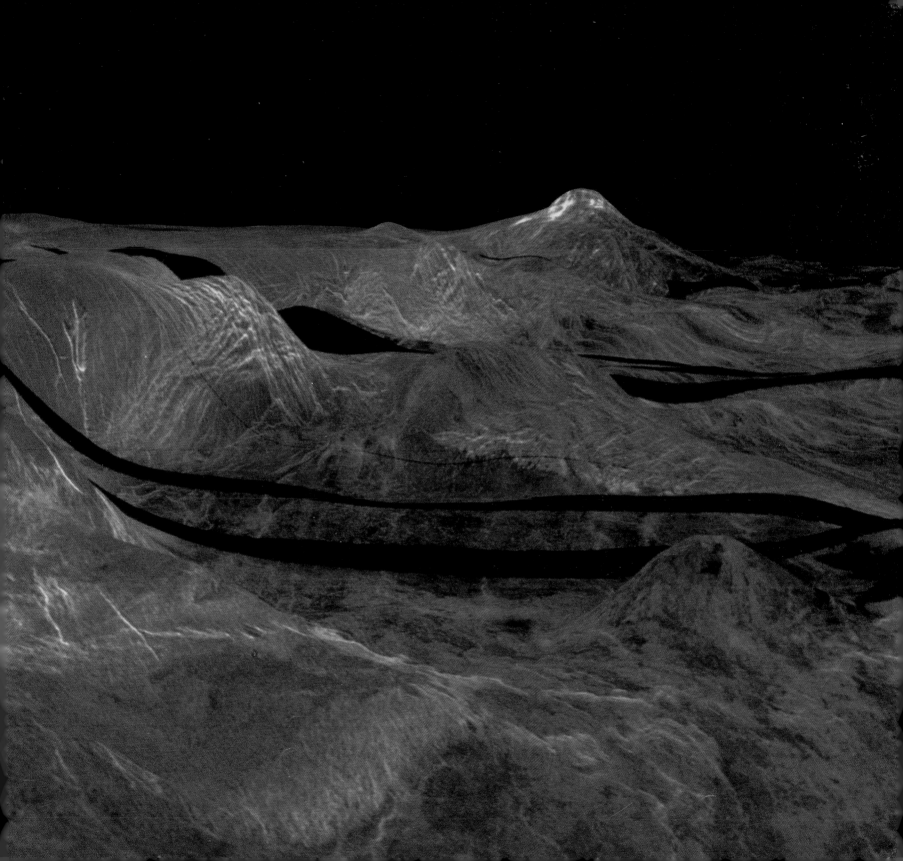

caused by all the water in Earth's oceans. About three-fourths of Earth's surface is covered by water. Earth's water and air, along with light and heat from the Sun, make it possible for plants and animals to live. As far as we know, Earth is the only planet on which life exists.

Even Earth's nearest neighbor, the Moon, has no life. The Moon orbits Earth just as Earth orbits the Sun. There is no water on the Moon and no air to breathe. Astronauts who landed there found a cratered landscape covered with gray, powdery dirt. The Moon is beautiful to look at from Earth, but you wouldn't want to live there!

The next planet beyond Earth is a red planet—*Mars*. Mars is about half the size of Earth and shines like a red star in the evening sky.

⇐
*Planet Earth rises above the Moon's horizon*

Mars has two very small moons that circle it. These moons are not round like Earth's moon. Instead, they are shaped like potatoes!

The surface of Mars has enormous volcanoes that no longer erupt. One volcano is fifteen miles high—three times taller than the highest mountain on Earth! Mars also has the largest canyon in the solar system. This canyon would stretch across the entire United States! There are also channels on Mars that look like dried-up riverbeds. Scientists think that Mars had liquid water on its surface at one time. Today, however, the only water on Mars is in its thin ice caps and the clouds of ice that float above its surface. Scientists estimate that all the water on Mars would fill only a small pond.

⇒
*A river channel meanders across the surface of Mars*
*Insets: Mars' Valles Marineris, the largest canyon in the solar system and Olympus Mons, a martian volcano that is three times taller than Mount Everest*

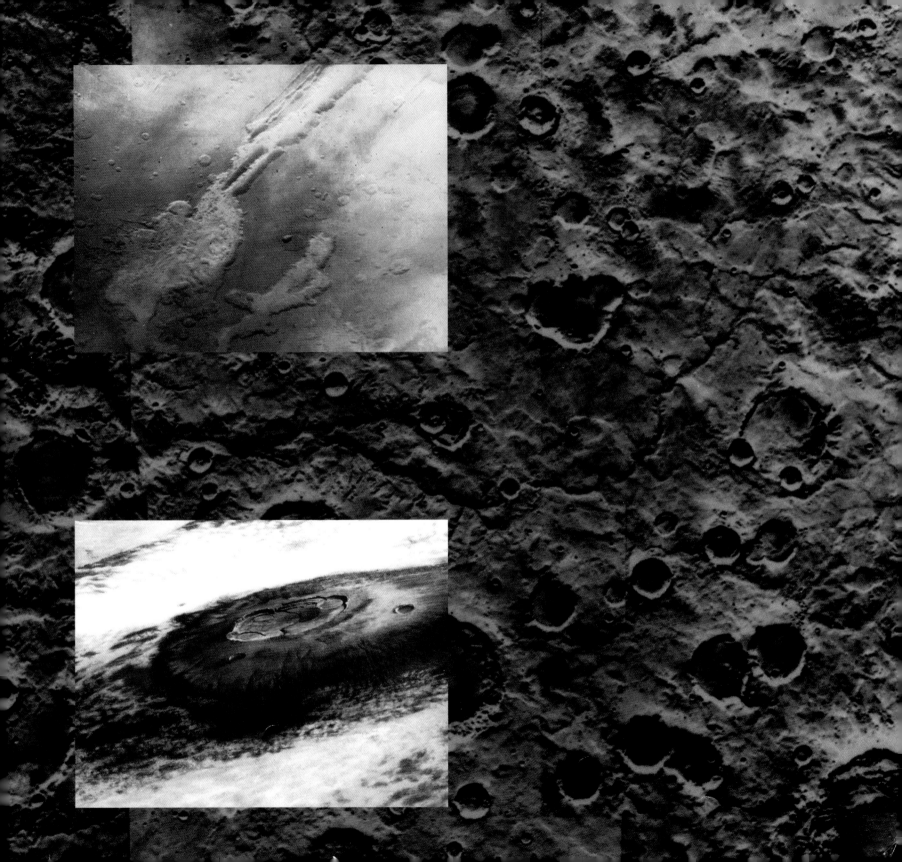

# INDEX

asteroids, 9, 19, 28

atmosphere, 10, 19

comets, 9, 28, 31

craters, 9, 19

Earth, 10, 13

falling stars, 19

Ganymede, 23

Halley's Comet, 31

Io, 23

Jovian planets, 20, 23, 27

Jupiter, 20, 23

Mars, 13-14

Mercury, 9, 23

meteors, 5

Moon, 13-14

Neptune, 27

outer planets, 20

Pluto, 28

Saturn, 23-24, 27

stars, 5-6, 19, 24, 31

Sun, 5-6, 9

Titan, 24

Triton, 27

Uranus, 27

Venus, 9-10

Comets orbit the Sun on regular schedules. The most famous, *Halley's Comet*, passes by Earth every seventy-six years. The last time it passed our planet was in 1984. If you missed seeing it, you'll have to wait until the year 2060 before you get another chance!

In the meantime, you can still enjoy gazing at the stars. Many of the small, twinkling stars you see are suns that are much larger and brighter than our own. Scientists suspect that many of these stars are orbited by planets.

Do you think there might be another planet somewhere in the universe that has oceans, plants, animals, and intelligent beings like the people on Earth? It's nice to think that this is possible. One day we may know for sure.

---

⇐
*The Andromeda galaxy, one of billions in our universe, contains billions of stars like the Sun*

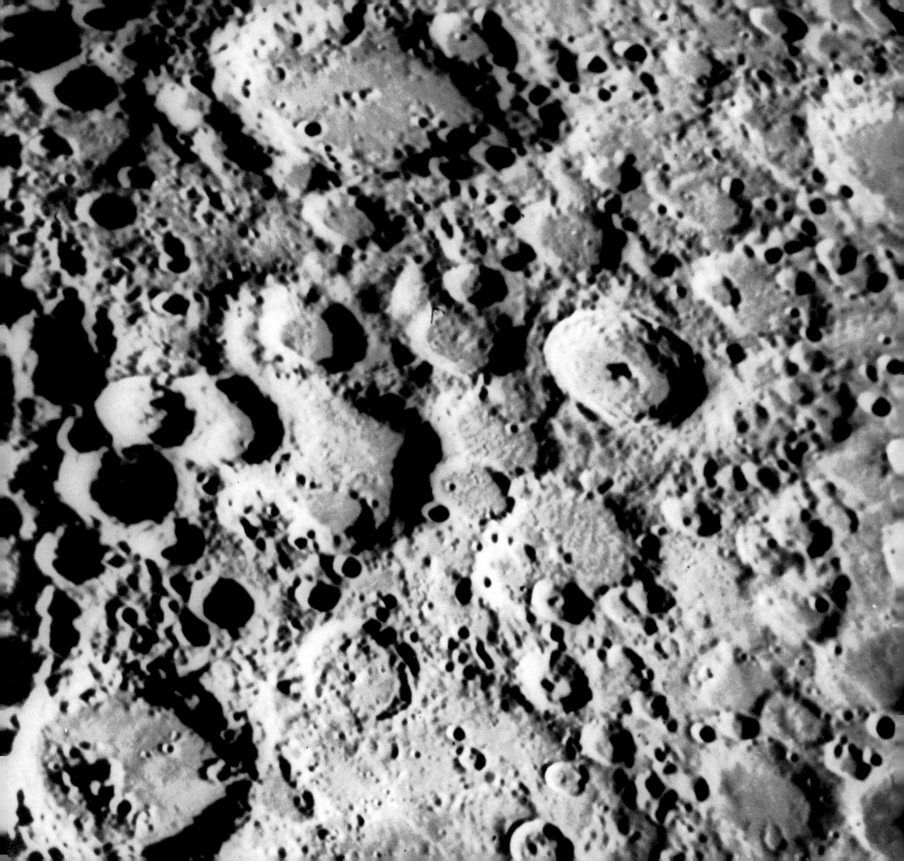

Before you reach the next planet in the solar system, you must pass through the *asteroid belt*. This is a wide band of rocks and metals that circle around the Sun. Most asteroids are the size of small rocks or pebbles, but others are as large as houses. The largest asteroid is 600 miles wide! Occasionally, large asteroids smash into one another and break apart into smaller pieces. Some of these small asteroids are pulled by gravity into the orbits of other planets.

On starry nights people say they see "falling stars." These streaks of light are really small pieces of asteroids, called *meteors*, burning up as they fall through Earth's atmosphere. Luckily, most meteors totally burn up before they crash to Earth's surface and form a crater.

---

⇐

*The craters on the Moon's surface were caused by falling asteroids*

The planets beyond the asteroid belt are called the outer planets. The first four outer planets, called the *Jovian planets*, are much larger than the other planets. They also lack the solid, rocky surfaces of the other planets. If you tried to land a spaceship on one of the Jovian planets, you would merely sink through layer after layer of thick, misty clouds.

The first of the Jovian planets is *Jupiter*. It is the largest planet in the solar system. In fact, it is larger than all the other planets put together! Jupiter consists of gases that move in bands of red, orange, and yellow across the surface. To some people Jupiter looks like a giant pizza floating in space. Strangely enough, the most obvious feature on Jupiter, called the *Giant Red*

⇒
Thick clouds cover Jupiter, one of the Jovian planets

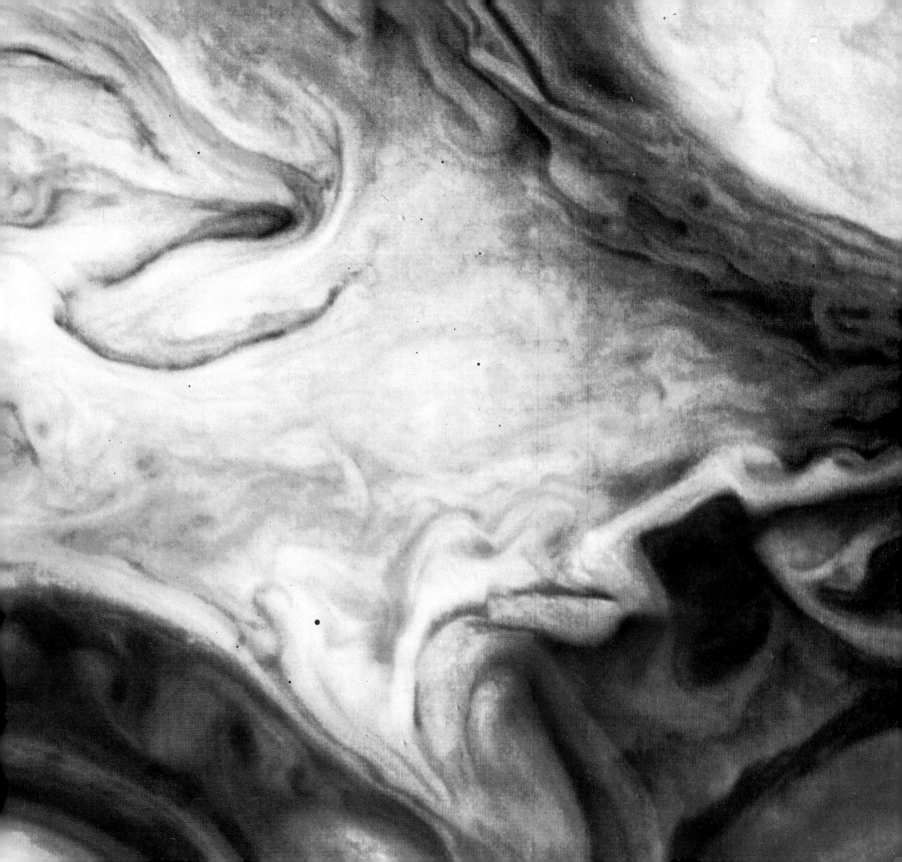

*Spot*, looks like an enormous piece of pepperoni! Scientists think the Giant Red Spot is a violent storm, much like a hurricane on Earth. This one storm is as wide as three Earth-sized planets set side by side!

Jupiter has sixteen moons. Most of these moons are small, but three are larger than Earth's moon. One, named *Ganymede*, is larger than the planet Mercury! The moon closest to Jupiter, *Io*, is a seething world of volcanoes and lava. Io has more active volcanoes than any other planet or moon in the solar system. One photo of Io shows eight volcanoes erupting at the same time!

Perhaps the most beautiful planet is *Saturn*. Like the other Jovian planets, Saturn is a large ball of gas. The planet is made up mostly of helium and hydrogen, so it is very light for its size. If you could find a big enough lake, Saturn would float!

Scientists have discovered twenty-two moons orbiting Saturn—more than any other planet. The biggest is *Titan*, which is 3,200 miles across—almost three times larger than the Earth's moon.

What distinguishes Saturn from the other planets is its set of distinct rings. Saturn looks like a large ball in the middle of a phonograph record! The rings, which consist of pieces of ice and dust, orbit Saturn much as its twenty-two moons do. Although the rings are over half a million miles across, they are less than a mile thick. In fact, you can see stars through them.

⇒

*Saturn and some of its twenty-two moons*

Beyond Saturn are three planets you cannot see without a telescope. They are very cold because they are so far from the Sun. The first of these distant planets—and the third of the gaseous Jovian planets—is *Uranus*. Astronomers have counted fifteen moons and ten thin rings circling this green planet. Uranus is different from all the other planets because it seems to be lying on its side as it orbits the Sun!

About one billion miles beyond Uranus is the fourth Jovian planet, bluish-green *Neptune*. Although Neptune is very far from Earth, astronomers have seen two of its moons. One of these, *Triton*, has active volcanoes. Triton is so cold, however, that its volcanoes spit out fountains of ice, not lava!

---

⇐
*Clockwise from upper left: Uranus, Neptune, and Triton*

*Pluto* is the ninth planet, and the farthest from the Sun. Pluto's great distance makes it difficult to study, but scientists have determined that it is the smallest planet. It is also the coldest planet in the solar system! Scientists think Pluto must be made of frozen gases. Pluto follows a very wide orbit around the Sun. It takes Earth one year to orbit the Sun, but it takes Pluto 248 years to complete one orbit!

In addition to the nine planets and millions of asteroids, the Sun is also orbited by *comets*. A comet is a huge ball of ice surrounded by rocks and dust. As a comet nears the Sun, some of the ice melts and turns into gas. The gas trails behind the ball of ice like a tail. A comet's "tail" can be millions of miles long.

---

⇒

*Pluto and its moon Charon are dimly lit by the distant Sun*

NYSSA PUBLIC LIBRARY
NYSSA, OREGON

22004